Chicago Public Library

REFERENCE

Form 178 rev. 1-94

TEACHING PRIMARY SCIENCE

Fibres and fabrics

John Bird and Ed Catherall

A Chelsea College Project sponsored by the Nuffield
Foundation and the Social Science Research Council

Published for Chelsea College, University of London,
by Macdonald Educational, London and Milwaukee

First published in Great Britain 1976 by
Macdonald Educational Ltd
Holywell House, Worship Street
London EC2A 2EN

Macdonald-Raintree Inc
205 W. Highland Avenue
Milwaukee, Wisconsin 53203

Reprinted 1977, 1978, 1980

© Chelsea College, University of London, 1976

ISBN 0 356 05076 9

Library of Congress Catalog Card Number
77-82983

Project team

Project organizer : John Bird

Team members :   Dorothy Diamond
                   Keith Geary
                   Don Plimmer
                   Ed Catherall

Evaluators :   Ted Johnston
                   Tom Robertson

Editor

Penny Butler
Macdonald Educational

with the assistance of
Nuffield Foundation Science Teaching Project
Publications Department

Made and printed by
Morrison & Gibb Ltd, London and Edinburgh

General preface

The books published under the series title Teaching Primary Science are the work of the College Curriculum Science Studies project. This project is sponsored jointly by the Nuffield Foundation and the Social Science Research Council. It aims to provide support and guidance to students who are about to teach science in primary schools.

Although the College Curriculum Science Studies materials have been produced with the student teacher very much in mind, we suggest that they will also be of use to teachers and to lecturers or advisers—in fact to anyone with an interest in primary school science. Hence this series of books.

Three main questions are considered important:

What is science?

Why teach science?

How does one teach science?

A very broad view is taken of teacher training. Training does not, and should not, stop once an in-service or college course has been completed, but can and does take place on a self-help basis in the classroom. In each context, however, we consider that it works best through the combined effects of:

1 Science Science activities studied practically at the teacher's level before use in class.

2 Children Observation of children's scientific activities and their responses to particular methods of teaching and class organization.

3 Teachers Consideration of the methods used by colleagues in the classroom.

4 Resources A study of materials used in the teaching of science.

5 Discussion and thought A critical consideration of the *what*, the *why* and the *how* of science teaching, on the basis of these experiences. This is particularly important because we feel that there is no one way of teaching any more than there is any one totally satisfactory solution to a scientific problem. It is a question of the individual teacher having to make the 'best' choice available to him in a particular situation.

To help with this choice there are, at frequent intervals, special points to consider; these are marked by a coloured tint. We hope that they will stimulate answers to such questions as 'How did this teacher approach a teaching problem? Did it work for him? Would it work for me? What have I done in a situation like that?' In this way the reader can look critically at his own experience and share it by discussion with colleagues.

All our books reflect this five-fold pattern of experiences, although there are differences of emphasis. For example, some lay more stress on particular science topics and others on teaching methods.

In addition, there is a lecturers' guide *Students, teachers and science* which deals specifically with different methods and approaches suitable for the college or in-service course in primary science but, like the other books in the series, it should be of use to students and teachers as well as to lecturers.

Contents

Introduction

Fine hair-like fibres like wool, cotton or flax are spun or twisted into yarn or thread which is woven or knitted into fabric.

This book concentrates on fabrics, but also deals with paper, which is similar to fabric because it too is made up of fibres, usually of plant origin. However, the constituent fibres of paper are compressed into flat sheets and are not spun or woven.

Because fibres and fabrics are so much part of everyday life, they are very appropriate materials with which to impart the fascination of science to children. The topic is very broad and can be pursued in many directions.

Chapter 1 discusses some of the problems of teaching and organizing this topic and considers how they might be overcome.

Each of Chapters 2–6 concentrates on one aspect of the topic. These chapters may be linked together, but each can be used independently as a resource dealing both with some aspect of *what* to teach and *how* to teach it.

The final two chapters broaden the scope of the book. Chapter 7 discusses activities with fibres and fabrics that are not generally thought of as scientific, and Chapter 8 considers further possibilities.

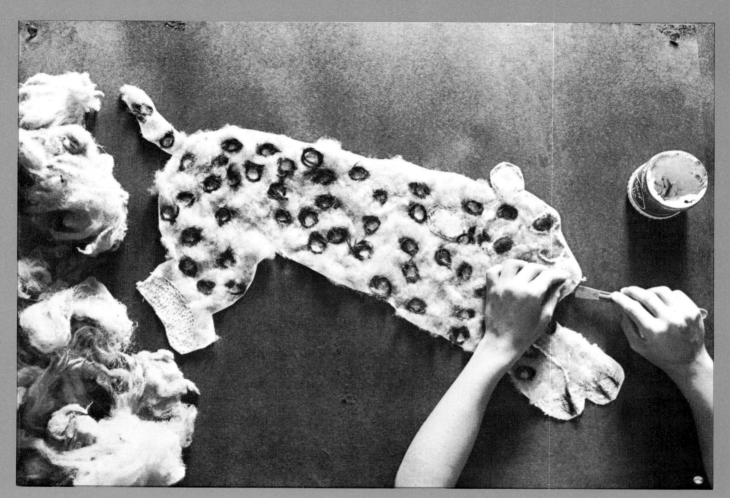

Making a collage with fibres

1 Materials and organization

Obtaining fibres and fabrics

Fibres   Here are some sources of supply.

Single threads   You can unravel these to get the fibres.

Human hair   is of course easy to obtain.

Wool   can be picked off fences in country areas where sheep are kept.

Raw and prepared wool   is stocked by spinning and weaving specialists and some educational suppliers.

Wool or nylon flock   may be bought in department stores. (Flock is used for stuffing toys.)

Cottonwool   may be bought cheaply in the high street.

Glass wool   may be bought from car repair shops.

Threads   can be obtained by unpicking fabrics or by buying some threads and yarns. The purchase may not be too expensive as a single roll or reel will go a long way.

Fishing shops stock single-strand nylon (or similar) fishing line.

You may be able to get hold of hemp, cotton, or sisal twine, which you can unravel, but it is generally being replaced by non-natural materials like the shiny polypropylenes.

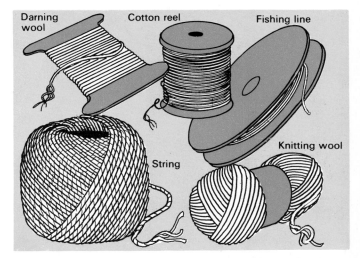

Fabrics   are not difficult to come by. Most schools have a box somewhere stuffed with fabric oddments. Discards from jumble sales are also to be found and will willingly, even eagerly, be donated to anyone prepared to take them away.

Children will certainly bring in fabrics from home if they are encouraged to do so. But don't be too insistent, or it won't be just discards that are brought in.

Shops and educational suppliers will sell fabrics, but this will be an expensive way of acquiring them.

See bibliography: 40, 41.

What other sources do you find useful?

Identification

It is helpful to identify and name some threads and fabrics, because discussion often leads to discovering the properties of the different types of fibres and fabrics.

Sometimes naming is not too difficult, although the modern practice of mixing different fibres in the same

thread or fabric may cause problems. Many clothes have labels which say what they are made of. Most sewing threads or knitting wools are similarly labelled.

This information will usually be enough to identify most of the materials you have. Some of the larger textile groups will supply free samples of named fibres and fabrics to schools.

See list of suppliers on page 40.

Identifying fibres

What's in a name? The variety of names for fibres and fabrics—especially those of man-made origin —is confusing. A name may be the material's common name or its trade name, or may refer to its basic chemical structure.

Here is a chart which may help to allot a fibre or fabric to its correct category. For example, the chart shows that Enkalon is the trade name for a form of nylon, whilst poplin is the common name for what is usually, but not always, a cotton weave. Books such as Marguerite Patten's *The Care of Fabrics* have more detailed information along these lines. Consult these if you want further help.

See bibliography: 38.

Fibre categories			
Chemical structure	Name of fibre	Types of fabric	Trade names
Protein	Silk	Chiffon	
	Wool	Tweed Twill	
Cellulose	Cotton	Poplin Denim Calico	
	Linen		
	Rayon		Vincel
Cellulose acetate	Di-acetate		Dicel
	Tri-acetate		Tricel
Nylon and polyester	Nylon and polyester		Enkalon Celon
Acrylics	Acrylics		Orlon Acrilan Courtelle Teklan

How to decide what to do

First comes the problem of what the children might do. There are many suggestions in the text, especially in Chapter 8. Consult also the bibliography on pages 38–39. You will have your own plans, but remember how creative children are, so talk to them and get their suggestions too.

Once you have got some ideas, you will have to decide which ones to take up and how best to adapt them to suit your needs as well as those of the children. Your choice will depend on the following factors.

Organization Do safety and the availability of materials limit what can be done?

Aims It may help to decide broadly on your aims. Do you want to encourage children's knowledge in the following ways?

Experiencing things Objects and phenomena like man-made fabrics or water.

Carrying on activities Things they can do or find out. These include not only scientific processes such as observation and experiment (see Chapters 2–6), but others less closely connected with science, like making or writing.

Developing ideas There are some which can be specially suitable for this topic, such as burning, safety or absorption.

Having fun Finally, an aim that is often forgotten: do you want the children to have fun?

A fuller description of how these aims might be developed is in Chapter 8. In any case, it is more a question of priorities than alternatives. For example, although you may place greater stress on *activities*, the children may still be dealing with *things* and *ideas*— and having fun too.

Interest You might also ask if an idea is likely to interest the children. If not, how can their interest be increased? Again, discussion with the children is important before you go ahead with your plans. Study how they react to various suggestions and consider what modifications or alternatives they suggest.

Age and ability   Will the children have the ability or the understanding to carry out these ideas with any satisfaction? If they haven't, it does not necessarily mean that an idea should be rejected. It may be that the level of difficulty may have to be lowered or that the teacher may have to give the children more help. Some guidance with this problem is given in Chapters 2–6.

Deciding how to do it

There are two main methods of organizing the work.

Class discussion   Discussion between the teacher and the whole class is very useful, especially for:

Giving instructions.
Obtaining suggestions from the children.
Providing children with information.
Developing understanding through shared ideas and experiences based on the children's practical investigations.

In *Teaching Elementary Science Through Investigation and Colloquium*, Lansdown, Blackwood and Branwein analyse class discussion, which they call colloquium, as a method for developing this sort of understanding, and stress its value. Clearly the key to a successful class discussion of this type is the active involvement of the children. In this book you are asked to consider if and how each of these methods in conjunction with displays set up by the teacher, workcards, recording and display by the children might be used for particular investigations.

See bibliography: 35.

Group work   This is another good way to organize the children for carrying out practical investigations.

How might you arrange the groups? Here are some possibilities.

1   The whole class is divided into groups each doing the same activity, say, the effect of water on fabrics.

2   The whole class is divided into groups each doing different activities.

3   One or two groups do investigations whilst the rest do something entirely different.

Discovery table   The teacher sets up activities on a discovery table and children try them out in odd moments.

What problems or advantages can you see in the use of each of these methods with a class? Which would you use? Have you thought of any other methods?

Beginning

How will you start? You will need to decide the following issues.

Starting points   Should the investigations be based on a broad integrated theme, or should they be much more narrowly and distinctively about science? See Chapters 2–6 and then Chapter 7 for examples of these alternative approaches.

Sequence   Might the work follow a logical sequence of development? For example, observation and classification of different fabrics might well come first because this would form a useful general introduction to the topic from which many further investigations might arise. Here also you might consider at what points and for what purposes you might use class discussion.

Grouping   How many groups will be involved? Are you intending to start in a small way with perhaps only one or two groups, or will you involve the whole class?

Controlling what is going on and the availability of materials are important here.

Position   Where should the groups be placed? Bear in mind, in particular, the need for:

Week 1	Fibres and fabrics	Elizabethan England
Class discussion	1 Old and new fibres and fabrics 2 Spinning and weaving methods	**Group work** Finishing frieze ⎫ Writing up topics ⎬ whole class
Request	Bring samples of fibres and fabrics by next week	

Week 2		
Display	Fibres and fabrics	**Group work** Writing up continues ⎫ 10 children
Class discussion	Differences between fibres and fabrics	Display begins ⎬

Group work	1 Water and fabrics: 3 children (see pages 16-19)
	2 Making spindles: 4 children (see page 32)
	3 Making a loom: 2 children (see page 33)
	4 Fabric printing: 6 children (see page 30)
	5 Research on sources of fibres and fabrics:
	   3 children (see page 34)
	6 Research on history of spinning and weaving:
	   2 children (see page 35)	

Request	Bring samples of vegetable material for dyeing	

Week 3		
Class discussion	Dyes and dyeing	Display complete
Group work	Including 10 children from previous topic who start work on: 7 Dyeing cloth: 4 children (see pages 28-30) 8 Research on textile industry: 3 children (see page 34) 9 Clothes through the ages: 3 children (see pages 34-35)	

Easy access to resources and services.
Avoiding unnecessary traffic through the classroom.
Class supervision, for example where safety is involved.

Furniture   Is it necessary to reorganize the classroom furniture? If so, how?

Speed   Will some activities be finished more quickly than others? In this case, what will those children do who have finished?

What will the rest do?   If only some children are involved, what will the others be doing at the same time? This is important because it is essential not to over-commit yourself. If the other children are doing something less demanding to the teacher but also less interesting to them they are liable to become restless.

Example   The diagram shown above is taken from notes written by a teacher about the first three weeks of a project on clothes and clothing which she did with a class of thirty children aged ten and eleven for an hour every Thursday afternoon. This topic ran side by side in the first two weeks with another on Elizabethan England which involved mainly research in books, writing, drawing and painting. No breakdown of how this was grouped is shown in the diagram.

Here science was linked to an integrated theme in a large-scale project.

The group activities by the children largely followed those suggested in this book (see page references on the diagram).

What main problems can you see arising from each week's work? How precisely would you plan each week's work?

2 Observation

Finding ways in which fibres and fabrics differ

You will need a large collection of different fibres and fabrics.

Looking   To begin with, work on your own. Search for those features that can be picked out by straightforward observation, by looking, pulling, stretching, and so on. These are the surface features. A magnifier might help here to show, for example, coarse structure.

Hidden features that can only be observed by means of special tests, such as burning, can be found later.

Using all the senses   The process of observation is not as simple as it may seem. It can involve *all* the senses, not just seeing, but feeling, smelling, tasting, hearing too. Try out all these senses, and see if there are any other observations you can make. Make a list. This is not only an aid to the memory; it also helps you to use words and so to clarify your observations.

Discussing   What one observes depends on one's knowledge and experience. People see things in different ways. Consult *Science and Society* by Michael Bassey for an interesting discussion of this point.

One's own observations can therefore be extended by discussion with others, including of course the teacher or lecturer. Through discussion one may get to know of observations one had never thought of making or of useful descriptive words. Frequently there will be differences of opinion about what one really is 'seeing' but this can only be a good thing as it is likely to lead to closer and more detailed observation and experiment. Discuss your observations and from them make a list of ways in which fibres and fabrics differ. (See bibliography: 32.)

Observing with children

Here is a brief extract from a discussion with James and Helen, aged eight.

Teacher: 'How many ways are these fabrics different?'
Helen: 'Well, these are plain, that's plain.'
James: 'This is all plain and that's patterny.'
Helen: 'This is got half torn off, and it's sort of got all threads. It's got dirty marks on it. This has got all writing over it.'
Teacher (plainly a little put off by the type of comments he is receiving): 'Have a closer look at these materials. What else can you see?'
Helen: 'Well, they've got different colours.'
Teacher: 'Which is the thicker of these?'
Helen: 'This one.'
Teacher: 'Can you see through this?'
Helen: 'Yes, you can see the little bits of holes. That hasn't got holes.'
Teacher: 'But surely it's got holes?'
Helen: 'They aren't actually holes.'
Teacher: 'What are they, then?'

What does this conversation show about the nature of observation and the value of discussion?

Would you have handled this discussion differently? If not, why not? If so, how?

Organization in class

When working with children the teacher can start by trying to find those features that can be observed directly without special tests. How can this best be organized in class? Taking into consideration your own experiences of observation, which combination of activities would you think most appropriate? Can you think of anything else? Here are some suggestions.

1 The teacher lays out different examples of fabrics on two desks at the side of the classroom. With them she displays a notice reading 'Who can find the most differences between these fabrics?' Groups of children visit the tables and explore the collection whenever time can be found.

2 As above. A book is provided as well, in which each child can record his comments and results.

3 The teacher lays the fabrics out at the front of the classroom and discusses with all the children the differences between the fabrics. The teacher hands the fabrics round to demonstrate what these might be, and lists the differences on the chalk board as they are discovered.

How acute are the children's observations? How are they different from your own?

What methods of organization did you use, and how successful were they?

Tape-recorded material or examples of children's work might help to illustrate your experience here.

3 Classification: burning

How to classify

Observation is closely linked with classification, that is, grouping fibres and fabrics according to different criteria. Classification can be done for its own sake, but it is more valuable when it is done for a specific purpose. The criteria will depend on the purpose. For example, children who are preparing a collage would collect fabrics which have features of colour or texture which are suitable for the design. Children who are studying farming might group fibres or fabrics according to origin—plant, animal, or man-made.

Methods
Ordering   Sorting materials according to a sequence, for example the extent to which they absorb water.

Grouping   Dividing materials into groups, for instance those that burn and those that do not burn.

A first attempt   Classify any or all of the following according to their superficial differences.

A collection of odd bits of fabrics
Various strings
Ropes (if available)
Threads
Human hair
Animal hair
Paper
Things made of paper
Various types of fibre such as cotton or wool according to their appearance under the microscope

What criteria might you use for classifying in each of these examples? There are obviously many differences between them.

For example:

String, rope and wool differ in ply.
Hair differs according to colour, thickness, or waviness.
Paper varies according to its thickness, smoothness, or absorbency.

This diagram shows some of the microscopic differences between wool and cotton.

See bibliography: 26, 39.

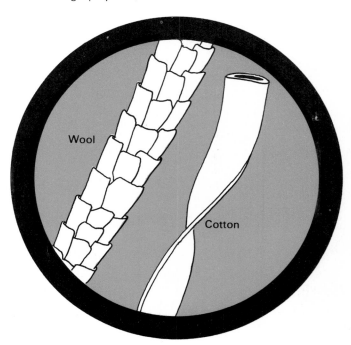

Wool

Cotton

Investigations

This classification is based on hidden differences between the materials which can only be revealed by special tests—in this case, burning. You will need:

A variety of fabrics and papers
Scissors
Tongs
15-cm length of wire
Candle
Nightlight
Matches
Large metal lid, tray or asbestos sheet
Two flat pieces of brick or tile
Large nail

Methods   Start by burning single threads from the fabrics. Then try narrow strips of fabric cut to convenient lengths, say 8 × 2 cm. The simplest method is to suspend a thread or strip over the flame. The material can be held with tongs, or it can be either threaded on to, or tied to, one end of the wire.

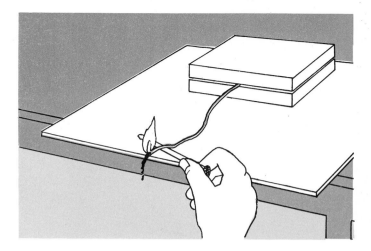

Another method is to fix one end of the thread between the two tiles (or pieces of brick). Then lay out the thread along the asbestos sheet with its other end hanging over the nail. Light this end, and as it burns move the nail slowly ahead of the flame.

Try out these methods. You may be able to make modifications of your own. Then decide which is the simplest and safest to use.

Types of investigation   Two types of test can be carried out:

Moving the flame close to the fabric but not applying it directly to it.

Trying to set the fabric alight and looking to see how fast it burns.

Questions to ask
Does the material change in any way when the flame is held *near* it? For example, does it melt or change colour?

Does the flame change colour when the material is held in it?

Does the material burn? If so, how fast?

If the material does burn, what changes occur? What colour is the flame?

	LINEN	NYLON NET	BROCADE	SATIN	TOWEL	COTTON	HAIR
Wear			Good				
The time it takes to die down	1 minute 36 seconds	2 seconds	1 second	11 seconds	2 minutes 49 seconds	18 seconds	2 seconds
Flares							
Smoulders							
Ash					Black	Black	
Melts							
Smell	Hair	Wood			Burning Paper		

Is there any melting?

Are smoke or fumes given off during burning?

Do the smoke and fumes smell?

Is there an ash left after burning?

What colour is the ash? Is it light and soft, or brittle?

Does a hard bead form after burning?

It might be helpful to record and tabulate the observations you can make in response to these questions and then discuss them. An example is given above. Refer back to the discussion of observation on pages 8–9.

Do you find recording and discussion helpful? If so, why?

Analysing your investigations

Finding hidden differences   These tests will have revealed differences that are not obvious by simple direct observation. Two fabrics may look very much the same, but they may react in very different ways to the burning test because of fundamental differences in structure and properties. Burning tests are a particularly good illustration that such hidden differences exist.

Classifying fabrics   You might try to classify the fabrics according to their potential danger. Two criteria might be used here. One is obviously the degree of flammability: the ease and extent to which they burn. The other is how much they liquefy as they burn, for hot burning liquid is obviously dangerous.

With your table of results (see opposite) work out a simple classification along these lines. You might start by ordering or grouping the fabrics according to each of these criteria. You might then combine the two criteria by means of a simple table or Venn diagram.

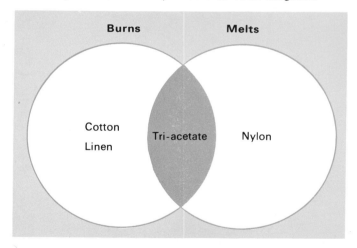

Identifying fabrics   Quite incidentally these tests also build up general knowledge about particular fabrics. This also involves classification, that is, sorting out observations according to whether they are typical of a given fabric. For example, cotton burns in a certain way and produces a different kind of ash from wool.

Consider this extract from a discussion with some eight-year-old children:

Teacher: 'What sort of ash have you produced with the wool, Helen?'
Helen: 'I dunno—it's very hard—it's something like coal.'
Teacher: 'What does it smell like?'
Steven: 'Burnt chips.'
Helen: 'Burnt supper.'
Teacher: 'Which is the one that burns the fastest?'
Steven: 'That one [the cotton].'
Teacher: 'What's the ash like?'
Helen: 'White and flimsy.'
Teacher: 'Try this stuff [a man-made fibre].'
(Pause)
Steven: 'Look at that!'
Helen: 'Er . . . it's all black and gooey.'
Teacher: 'Which of these materials is the most dangerous?'
Helen: 'This one [the man-made fibre], 'cause it's all gooey and it burns.'

How many criteria which you could use for classifying fabrics can you find in this conversation?

Language *Teaching Elementary Science Through Investigation and Colloquium* stresses the importance of language in science. The authors quote Vygotsky's view that: 'Thought is not merely expressed in words: it comes into existence through them.' The children here are articulating a number of familiar words and phrases, but in a new and unfamiliar context.

See bibliography: 35.

Look carefully at this conversation and consider how such free use of language may here lead to fresh understanding. What advantage can you foresee if the teacher were to introduce more conventional words such as 'viscous' instead of 'gooey'? Or might this be a positive disadvantage?

What do you think?

An example of a key The key below can be used to help identify fabrics on the basis of burning tests. It is adapted from *Science Session Spring 1974* (see bibliography: 7).

An example of another key, extending to other criteria apart from those based on burning, can be found in Science 5/13 *Coloured things* (see bibliography: 23).

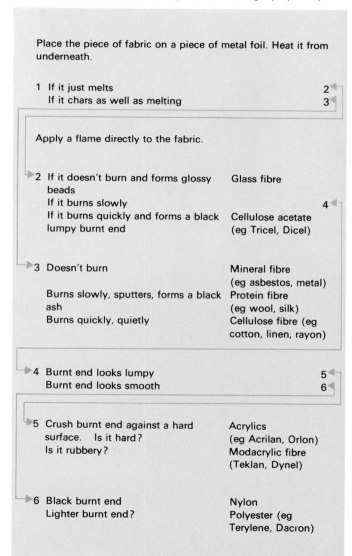

Place the piece of fabric on a piece of metal foil. Heat it from underneath.

1 If it just melts	2
If it chars as well as melting	3

Apply a flame directly to the fabric.

2 If it doesn't burn and forms glossy beads	Glass fibre
If it burns slowly	4
If it burns quickly and forms a black lumpy burnt end	Cellulose acetate (eg Tricel, Dicel)
3 Doesn't burn	Mineral fibre (eg asbestos, metal)
Burns slowly, sputters, forms a black ash	Protein fibre (eg wool, silk)
Burns quickly, quietly	Cellulose fibre (eg cotton, linen, rayon)
4 Burnt end looks lumpy	5
Burnt end looks smooth	6
5 Crush burnt end against a hard surface. Is it hard?	Acrylics (eg Acrilan, Orlon)
Is it rubbery?	Modacrylic fibre (Teklan, Dynel)
6 Black burnt end	Nylon
Lighter burnt end?	Polyester (eg Terylene, Dacron)

Safety in the classroom

Safety is of paramount importance in the classroom. Burning fabrics are safety hazards but they can be reduced to a minimum by taking sensible precautions.

Here are some specially relevant points:

The source of flame   Which of a candle, taper, or a nightlight is likely to be safest? (See bibliography: 8.)

Testing surface   Do the tests above an asbestos sheet or a large tin lid.

The type of holder   Obviously this should be long enough to keep the children's hands well away from the flame. It should not be a good conductor of heat and it should be at least long enough so that it does not get too hot to hold.

Positioning   The burning tests should not be carried out anywhere near materials such as papers, hair or trailing clothes which could be set alight. They should be done on a side bench or table well away from possible sources of risk.

Control   All these points raise the question of whether the tests should be done by the children themselves (perhaps in a small group) or as part of a demonstration/discussion by the teacher with the whole class.

Consider the best ways to organize these tests in your class bearing the question of safety in mind.

See also *Candles* in this series (see bibliography: 8).

Why teach science?   Children studying fibres and fabrics may become more safety-conscious and more aware of the reasons for sensible safety precautions in the home. One reason for teaching science, therefore, is that it can help us all to cope with the problems of day-to-day living.

Some investigations   These may help children to understand about the danger of fire and what to do when fire breaks out.

1   What treatments make fabrics less susceptible to burning, for example borax solution. (See bibliography: 20, 21.)

2   The effect of excluding air from burning fabrics and other materials. (See bibliography: 8.)

3   The ease with which long trailing threads and fabrics burn. For example, which burns more easily, a long strip of cotton, or a similar strip folded tightly into a ball?

How much do you find that children are aware of the simple rules governing fire safety, particularly about clothes? How far do you feel that they are helped by investigations like those above?

Classification in the classroom

Consider the following ways of helping children to classify in the classroom. Here we are talking about classification in wider areas than the one described in this chapter.

The teacher displays the materials under prepared headings.

The teacher gives the children a collection of materials and asks them to lay them out under prepared headings.

The teacher and children discuss ways in which the materials can be classified.

The children tabulate their results using a scheme prepared by the teacher.

The children use a key, perhaps pictorial, to identify different fabrics.

The children do investigations to help them classify.

Obviously ability, age differences and your objectives need to be taken into consideration here. For example, are you aiming to teach children *how* to classify? If so, a flexible approach using many different.criteria might be best. On the other hand, if you simply want to teach the children *one* method of classification you may adopt an approach that will be more straightforward and more direct.

4 Explanation: water and fabrics

An explanation is an interpretation that we place on our observations. This is an imaginative process, for there can be many explanations of an observation. These will vary according to the age, experience and ability of the observer.

Explanations should not be confused with ideas. When we explain something we are referring to a specific event. In doing so we make use of general ideas based on our past experience. Such general ideas can be applied in many situations.

You will need:

Various fabrics
Various papers
Drinking straws
Glass dropper
Candle
Oil
Liquid furniture polish
Vaseline
Liquid detergent
Stop clock or watch with seconds hand
Pin

1 Place drops of water on different fabrics and papers. It will help your comparison if the drops are all the same size.

Time the rate at which the drops are absorbed. Do new and used fabrics absorb drops at different rates? You may find that a drop is not absorbed even though the fabric is quite obviously porous. You may also notice that drops take on different shapes on different surfaces.

How would you explain this? How would you account for the resistance of some fabrics to water?

2 What happens to the resistance of fabrics that normally absorb water when you treat them with substances such as candle wax (which you can rub on), oil, or liquid furniture polish?

Work out a test that will clearly show what the effects are.

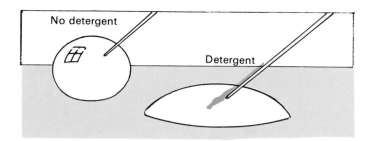

3 What is the effect of detergent on the shape of drops of water and their rate of absorption?

Apply detergent to drops with the end of a pin. What happens? How do you explain this?

Children's explanations

Consider this example. Mark (aged ten) had rubbed some wax on a fabric and when drops were placed on the treated area they became more rounded.

Mark: 'Hey, it doesn't go through. Is that air bubbles?'
Teacher: 'Now, you say "Is that air bubbles?" What do you think, Karen? Are they air bubbles?'
Karen (not really convinced, yet eager to please): 'I *think* so.'
Teacher: 'But why have they become more rounded?'
Mark: 'Perhaps the wax is pushing air into the water.'

Mark was probably basing his explanation on two specific ideas: first that drops are bubbles and have air in a central space, and secondly that bubbles will expand and become more rounded when they fill with air. Surprisingly even top juniors find the distinction between small drops and bubbles difficult.

Mark's explanation was of course only one of many possible ones.

Refer back to the activities that you tried out.

If you were working as a group what explanations did you give?

Analysing your explanations

When you are analysing your own explanations you may find the following ideas useful.

Absorption   Water enters the spaces between fibres and threads and/or is absorbed into the fibre.

Repulsion   Some fibres, such as Terylene, are water hating (hydrophobic) as are some substances like wax.

Attraction   Some fabrics, like cotton, are water loving (hydrophilic).

Do you find these ideas useful? If not, this may be because they are too simple to satisfy you. If you do, turn to 'Abstractions' on page 19.

Some further investigations

You will need:

Various fabrics
Various papers
String and threads of different types
Dessertspoon
Jam jar
Glass dropper
Measuring cylinder
Stop clock or watch with seconds hand

4 Cut strips of fabric sized about 10 × 1 cm. Place one short end of each strip in water. How far, if at all, has water travelled along the strips in a period of, say, half an hour?

Try the same method with string or thread.

Explain. In doing so you might use the idea of:

Capillarity   The tendency of water to pass along narrow tubes or grooves.

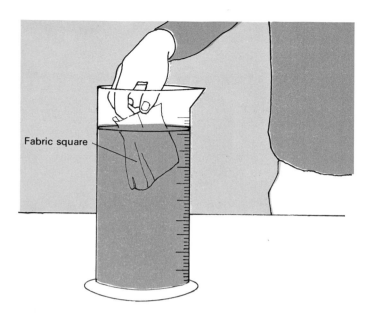

Fabric square

5   Cut out fabric squares of equal sizes and carefully insert each into a measuring cylinder full of water. Let them soak for some time and then lift them clear of the water and let any excess water drop off. How much water has each piece absorbed? Explain.

6   Do some fabrics actually let water through? Tie some fabrics across the mouths of jam jars in such a way that they sag in the middle. Place equal amounts of water (say dessertspoonfuls) on each piece of fabric. Does the water pass through the fabric and drop to the bottom of the jar? If so, is the rate for each fabric different?

How might you explain the difference? This explanation might make use of several of the ideas mentioned already, as well as the following one:

Permeability   The degree to which water (and other materials) passes through things—in this case, fabrics. This may be extended to include the idea of the 'permeability' of fabrics to air and light.

Washing

Work out some tests to find out the following.

How easy fabrics are to clean   Rub 'equal' amounts of one or several kinds of 'dirt' on some fabrics and then try to wash them off.

What factors help in washing?   How effective are hot and cold water, detergent, soap in removing dirt from any given fabric?

Here are two helpful ideas:

Solution, suspension   When things are dissolved they break up into very small particles which mix with the particles of the liquid they have dissolved into. In a solution the particles are too small to be separated from the water by filtering. In a suspension they are larger and can sometimes be filtered off.

The connection between washing and shrinkage   See Nuffield Secondary Science *Theme 7* for one suggestion about how to do this. Also consult *Science in the Home: Fabrics and Laundry Work* by Lilian Gawthorpe for an explanation of why fabrics sometimes shrink when they absorb water.

See bibliography: 13, 37.

Types of explanation

If explanations are examined and recorded carefully they can be classified into many different types. Here are some examples.

Looking for likenesses   Helen, aged eight, had just touched some fabric underneath where it was wet and the water had come through.

Teacher : 'Now why was that ?'
Helen : 'Well, 'cause it's ever so wet, and if you touched it it's still wet under there—because we had a tent what got all wet. We had hundreds of patches, we got very wet.'

Here Helen was explaining her observation by referring to another similar situation from her own experience. It was clear that she had understood something, since otherwise she would have chosen a less relevant example. Explanation by comparison is very common among primary school children.

Consider for each of the investigations in this chapter some 'like situations' which might help you to understand what happens. For example, in some ways the flow of water along strips of fabric is like oil flowing up through a wick, or water rising through a pot of soil.

Linking different observations together
Teacher : 'Now why do you think the water goes through ?'
James (aged eight) : 'Because there's holes in it and it's not thick.'

James was linking his observation that water went through with two possible contributory factors, and it is obvious that he had some understanding. A much younger child might have chosen colour as a relevant factor.

General ideas   Here the explanation is crystallized in a word representing a general idea which has already been built up by previous observations.

Graham (aged ten) : 'There was a puddle of water on it but it disappeared.'
Teacher : 'But has it actually dropped through the fabric ?'
Graham : 'No, it's dissolved.'

Graham's general idea that the water has 'dissolved' seems to be inappropriate, but is it really ?

Abstractions   The above types of explanation make relationships between things that can be directly observed. But they do nothing to explain *how* these factors are linked together. To do this one has to go beyond what can be observed and think in terms of abstractions.

Consider Mark's explanation of why drops on a waxed surface become more rounded (page 17).

An abstract or theoretical explanation of his investigation might be as follows. A drop of water is made up of tiny particles called molecules, all of which are attracted to each other. This attraction also produces a surface film which acts as a kind of envelope and helps the drop to retain its shape. The surface on which a drop rests also exerts a pull (called adhesion) on the drop. Some surfaces like wax exert less pull than others and the drop is therefore more rounded.

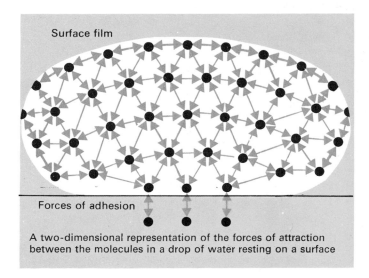

A two-dimensional representation of the forces of attraction between the molecules in a drop of water resting on a surface

Working with children

Now start one or more of the suggestions for investigation in this chapter and bring in further extension activities as and when required. Encourage discussion and help the children to give explanations in their own way.

What variety of explanations did the children give? What form did these take?

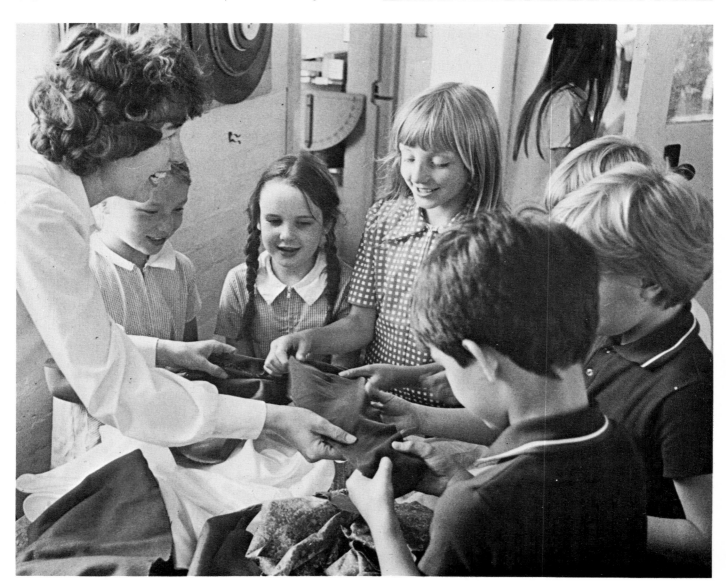

5 Experiment: fair tests

Sometimes an explanation can be put to the test. We suggest that *if* an explanation is correct, *then* this will happen, and we construct an experiment accordingly. An 'if . . . then' statement may be called a hypothesis. At least one part of Mark's explanation (page 17) could be tested quite simply.

Teacher: 'How would you know if they *were* air bubbles?'
Mark: 'Pop them.' (This was the hypothesis: *if* they were air bubbles *then* they would pop.)
Teacher: 'Try a pencil.' (Karen tries to pop them.)
Mark: 'It's not air.'

Refer to the investigations on pages 16–17. What hypotheses did *you* use and how did you test them?

When we test an explanation we try to make the test as fair as we can. There may be many possibilities but we ensure that we test one at a time, changing it around to see what effect it has, but keeping all others equal.

Children testing

Mark was working out a waxed paper test with the teacher.

Teacher: 'How can we make this fair?'
Mark: 'Make sure the paper was the same.'
Teacher: 'What else?'
Mark: 'You don't put wax on this side.' (He was comparing treated and untreated areas.)
Teacher: 'Anything about the water?'
Mark: 'Make sure it's the same.'

Teacher: 'What do you mean, the same?'
Mark: 'Well, the same quantity and . . . um . . . the same water from the same tin.' (An imaginative suggestion.)

What hypothesis was Mark probably putting to the test? What factors was he hoping to hold constant to make the test 'fair'? Refer back to your own investigations. How did you ensure they were 'fair'?

Consult Science 5/13, *With objectives in mind*, for a discussion of fair tests (see bibliography: 28).

Are fair tests always possible?

Is it always possible to carry out a test as fair as Mark's was likely to be? Sometimes it may not be so easy to hold all factors, except those under test, equal.

Consider this conversation with Graham who wanted to find out which of cotton, wool and a man-made fabric, absorbs water most quickly. He had already tested the wool.

Teacher: 'Has it gone through? Yes, so that was . . .
Graham: 'Two minutes fifty seconds.'
Teacher: 'For the water to soak into, but not through, this cotton. Now what material is this dark stuff?'
Graham: 'Man-made. It shouldn't take long 'cause it's thin.' (Here he has introduced an explanation of why some fabrics absorb water more quickly than others.)
Teacher: 'Why do you say that?'
Graham: 'Well, this cotton's thinner than the wool and that didn't take so long.'
Teacher: 'Is this cotton thinner than the man-made?'

Graham: 'No, it's thicker.'
Teacher (referring to the man-made fabric): 'So you are in fact saying that it'll go in more quickly?'
Graham: 'Yes.'

If Graham were to test the notion that thin fabrics absorb water more quickly than others, what *other* factors would he have to take into account when constructing a fair test? Do you think this would be easy to do? Some tests which are carried out do seem to be unfair. Do they, however, have some value?

Constructing fair tests on wear and crease resistance

Consider how you might investigate the degree to which two different fabrics (say nylon and cotton) become worn by abrasion, in this case using sandpaper.

You will need:

Sandpaper (coarse, medium and fine)
Wood blocks
Drawing pins
Bricks
Fabrics

First, what is your hypothesis? Secondly, what factors will you have to try to hold equal in order to ensure that the test is fair? One factor is the rate at which the sandpaper is used to rub the fabric, but there are at least two others. Can you think what they are?

Now construct the wear test. Use the materials suggested or any others that you think will be more effective.

What were your results? To what extent did they support your hypothesis?

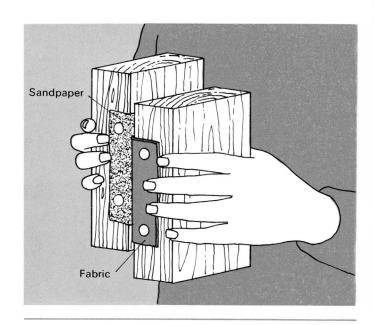

Sandpaper

Fabric

Helping the children with testing

Many children, especially younger ones, find fair tests difficult to work out. In the first place they may not be aware of all the factors that need to be taken into consideration. In this respect they are the same as many adults. In the second place they often have difficulty in dealing with different factors at once so as to hold them all constant except the one under test.

But if this is so, and it will certainly often be true of infants and lower juniors, what is the teacher to do? Perhaps she should have a discussion with the children in which she carefully guides them through the various stages of the experiment, where necessary showing them which factors need to be considered.

Why not try this method, perhaps in the form of a class discussion? How successful do you think it is?

6 Measuring

Why measure?

Teachers encourage children to measure for many different reasons. Here are some.

Because it is fun   It is worth doing for its own sake.

Because it is expected   by both children and teachers.

To provide raw information   Certain regularities and irregularities can be observed which might set you thinking about 'why' and 'how'.

To solve a specific problem   One might wish to raise the level of accuracy of the tests one applies to classify fibres and fabrics, in order to have a definite standard of comparison. For example, rather than saying that fabrics are *very*, *moderately*, *partly* absorbent, they could be classified according to the length of time they take to absorb a standard amount of water. As a result the data produced might give rise to questions and explanations, which could be tested by measurement. Thus this method might be used to test the suggestion that treatment with various substances changes the rate of absorbency.

The relationship between measurement and classification, testing and explanation could be drawn as follows.

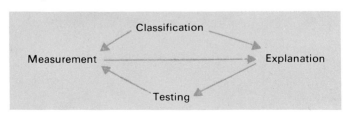

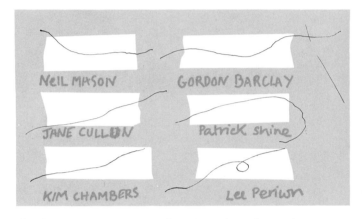

A classroom example   Here is a discussion between a teacher and a group of top juniors. The children are testing the breaking point of human hairs, in this particular case, one of their teacher's.

Lee : 'Seventy, eighty . . .' (grammes)
Teacher : 'Snapped?'
All : 'Yes.'
Teacher : 'Ninety, wasn't it? Ninety.'
Gillian : 'Do you think it's got anything to do with age—if you don't mind me saying so?'
(Laughter)
Teacher : 'Thank you! You're very nice! Well, I think it *might* do.'
Gillian : 'Shall I get one of mine out?'
Teacher : 'Well, we could try one of Pete's. He's got longer hair than I have, I think.'
Gillian : 'The boys we've done are Neil, Patrick and Gordon, and those all snapped at fifty.'
Teacher : 'Girls might have stronger hair.'
Gordon : 'Yes, but what about Sir's? His didn't.'
Teacher : 'Yes, but I'm an old man.'
Brenda : 'No you're not—not all *that* old.'

Name	COLOUR	LENGTH	CURLY OR STRAIGHT	UNDER MICROSCOPE	WEIGHT
Neil	Fair	SHORT	CURLY	LOOKS LIKE FIBRE FROM NYLON	50 Grammes
Lee	Fair	SHort	wavy	LOOKS LIKE GOLDEN THREAD	50 g
Gillian	LIGHT BROWN	Long	straight	LOOKS LIKE SEE THROUGH HOSE	100 g
Brenda	light Brown	Long	wavy	LOOK LIKE GUITAR STRING	60
Jane	Fair	short	straight	LOOKS LIKE VIOLIN STRING	60 g
Kim	Brown	long	straight	LOOKS LIKE	50 g

A child's table showing characteristics and breaking point of hair

Which of the general purposes of measurement do you think this exercise served? How do you think that this teacher handled the discussion? Which, or which combination, of these purposes do you think most important for your pupils?

Investigations

You will need:

Scissors
Ten large washers
Two 30-cm lengths of 1·25 cm ($\frac{1}{2}$ in) dowelling
Two sets of ten 100 g weights and hook
A wooden board (say 15 cm × 2 cm × 100 cm)
A piece of wire (say 10 cm long)
Two small nails
Sticky tape
Pencil
Ruler
Some human hairs from different heads
A selection of threads, such as nylon, fishing line, sewing thread, woollen thread
Papers, such as newsprint, writing paper
Fabrics

Testing the strength and stretch of hair fibres   Stick one end of a hair on the board with sticky tape. The hair will not come away if it is curled round under the sticky tape. Loop the other end of the hair round a hook made from the wire, and secure the loop by crossing the hair under a sandwich of sticky tape (see diagram). Measure the length of hair between, say, the point of attachment to the board and the hook, and then add equal weights (the washers), repeating your measurement each time.

Does the hair stretch? How many weights are needed before it breaks?

You might try again with a hair from the same source. How do your results compare?

Try comparative tests between hairs of different types (for example thin and thick, long and short). Perhaps also try animal hairs such as wool. What is the effect of soaking the hair in water on strength and/or stretch?

Testing the stretch and elasticity of different threads   Try to use threads of approximately similar thicknesses. Cut the threads into equal lengths, say 50 cm. Make sure that you can identify each thread. Tie the thread to a suitable fixing point, such as a nail, and tie the other end to a hook to carry the weights. Tie a thin wire pointer to the hook. Gradually add equal weights to the hook and measure the length each time.

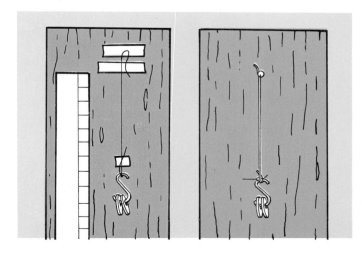

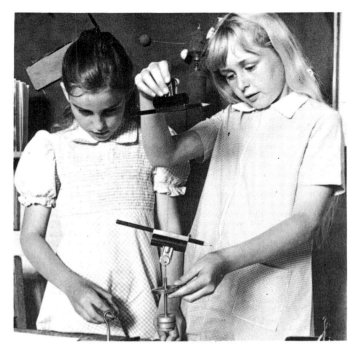

Write down the weights and lengths and then make a graph of the total increase in weight and the increases in length which went with them.

Now remove the weights one by one and record the lengths of the threads. Does the thread return to its original length each time? Plot this on the graph.

An elastic substance is one that is able to return to its original shape when the 'deforming' force is removed.

Repeat this stretching experiment with threads of different materials.

Testing the breaking point of threads
Start by testing one thread in the way demonstrated above.

Now twist two of these threads round each other. This is called a two-ply thread. Test to breaking point again. How close was the breaking point of the two ply to twice the breaking point of a single thread? Try two ply of different materials.

Then try a three-ply thread. How does this compare with treble the breaking point?

Graph the number of ply against the breaking point load, and put in the result you expected. What do you notice?

Try this test the other way round by trying four-ply wool and gradually untwisting it to three, two and one ply as you test to breaking point. An interesting account of investigations on the strength of fishing line can be found in *Science Teaching for ROSLA* (see bibliography: 10).

Testing the strength of paper
Use pieces of paper about 3 cm wide. Use sticky tape to attach the paper to the dowelling (see diagram). You can attach the other end in the same way, or hold it. Add weights until the paper breaks.

Repeat the experiment with different types of paper. What is the effect of soaking each of these papers in water? How do your results alter when you use the same paper but different widths? What effect does the length of paper have?

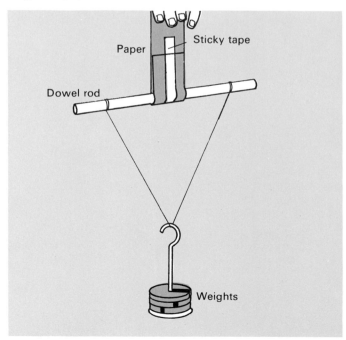

Analysing your investigations

Is accuracy important? Measurement involves comparison against some form of standard unit. In the tests above the amount of stretch was measured against arbitrary units of weight such as numbers of washers or standard units such as 100 g weights.

The units do not have to be as accurate as this, but it would be difficult to do these tests with less accurate ones.

However, if you want to group fabrics roughly and quickly according to the degree of stretch, it is enough to order them according to whether they are *very, quite, not very* stretchy, or to call the amount of stretch *little, some, a lot*. Here the units are broad and arbitrary, but quite adequate for some comparisons.

Thus the accuracy of the method and the units used will depend on what the measurement is for and the age of the children; there is no merit in making it more complicated than it needs be. Finally, you should remember that basically no measurement can be completely accurate; it is rather a question of degree of accuracy.

Ought tests to be reliable? Do measurements of the same things give consistently similar results?

If not, they cannot carry much conviction and the method needs to be scrutinized carefully to see how or why it produces such inconsistent results.

Granted that the method of measurement does prove reliable, however, another question must be asked.

Are the results valid? Do they satisfactorily answer the problem they were to 'solve'? It is essential to be quite clear about what the problem is in the first place, and secondly, to take into account the way in which results from a test may vary. For example, apparently identical threads may have different breaking points. Therefore if the question asked is what is the breaking strain of a certain thread, it is not valid to base your answer on only *one* measurement.

Some children prepared some strips of different kinds of paper so that they could compare the breaking points. Here they are. What might you say about the tests' reliability and validity?

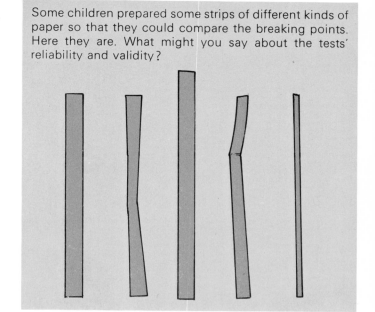

Adapting measurement to the children's abilities

The methods of measurement need to be tailored with care to the age and ability of the children. Manipulative problems are bound to be a limiting factor. You should also ask if the children appreciate the significance of what they are doing. For example, do they realize what may seem obvious to an adult, that the tension in a thread or fibre is equal to the weight hung on the end?

Methods of measuring the stretch of fabrics
1   Pulling as hard as possible on strips of fabric and ordering them according to the degree of stretch.

2   As no. 1, but using strips of similar length and width and measuring the increase in length against a scale.

3   As no. 2, but stretching each fabric using a single known weight.

4   As no. 3, but for each fabric adding a succession of equal weights and measuring the change in length each time.

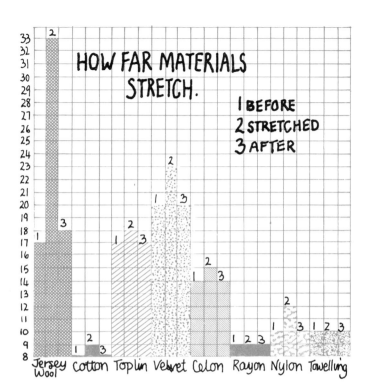

HOW FAR MATERIALS STRETCH.

1 BEFORE
2 STRETCHED
3 AFTER

Jersey Wool Cotton Toplin Velvet Celon Rayon Nylon Towelling

These methods are not exactly comparable, as they are not all for measuring quite the same things. However, which do you think is in a form most appropriate to your pupils? Look back at the other suggestions for measurements on pages 24–25. How can you adapt these to the abilities of your pupils?

Consider if and how your pupils might measure the rate of burning of different fibres and fabrics.

Measurement in class

Here are some methods of organizing measurement in class.

Work out precisely beforehand which tests the children should do.

As above, but leave the children some choice.

Only do tests when they arise from discussion with the children.

Prepare the apparatus for the children.

Leave the children to construct the apparatus themselves.

Detail on workcards how the children should do the measurements.

Allow the children scope to work out the methods of measurement themselves.

Discuss the children's work critically although sympathetically.

Accept the children's results without excessive discussion and criticism and record and display them appropriately.

Which combination of methods would you use? How would your choice be affected by the age or ability of the children?

7 Broad themes: making, creating and studying

So far the investigations have been rather narrowly scientific. They may capture the children's interest and involvement, but they may not. It is sometimes argued that it is better for the science to emerge from activities which have a broader scope, linking it for example with creative work, geography or history.

The following suggestions attempt to fit in with this approach. Links are suggested between these and some activities in previous chapters.

Dyeing fabrics

You will need:

Alum
Cream of tartar
Pieces of cotton, wool, and man-made materials, preferably white
Glass dropper
Drinks, eg coffee, tea, cocoa
Inks
Lichens
Leaves, eg red cabbage, onion bulbs
Berries, eg privet, blackberry, elder
Roots, eg dandelion, beetroot, red cabbage
Flowers, eg ragwort
Paints
Teaspoon, tablespoon
Two small saucepans
Wooden spoon for stirring.
Source of heat, such as an electric ring (careful!)

Extracting dyes   Many dyes can be extracted from natural materials like lichens, berries and roots. For useful sources see bibliography: 11, 16, 31.

Make the dye by soaking some plant material in water overnight and then boiling for forty minutes or more in a saucepan. Then pour the liquid into another saucepan.

Testing dyes

1   Test the effect of various paints, dyes, inks, food materials on some fabrics. Place drops of each of your samples on each type of fabric. Allow to dry and then try to wash them out of the fabric. Which dyes 'take' and which are washed out? By what process do you think that the dye is removed?

2   If a dye does not seem to 'take' on a fabric, try a more protracted treatment. Place the fabric in the dye in a saucepan and simmer for one hour. Remove the dyed fabric and dry it; then wash it. Does the dye hold fast?

3   If the dye still does not hold fast, try a mordant. This is a chemical which combines with the dye so that it holds fast to the fabric. A good one is alum. Add a tablespoonful of alum to four pints of water. Add half a teaspoonful of cream of tartar. Stir until the water clears. Heat the water and when it is warm add the fabric. Heat almost to boiling and then leave to simmer for half an hour. Remove the fabric and squeeze out as much water as possible. Then dye the fabric as in no. 2. Does the dye hold fast now? Refer back to the investigations involving absorption on pages 16–18.

See bibliography: 11, 16, 31.

Organization   The extraction of dyes and the dyeing and mordant processes present certain problems of organization, notably safety and the long time intervals involved. The children who do such projects will be older juniors with whom there are fewer problems of supervision and control. Otherwise, it might be better if the teacher does the boiling himself outside class hours.

How could or how did you do it with your class?

Separating colours   Try to separate the different coloured constituents of some dyes and inks as they soak in solution through paper or fabrics. An excellent description of how this might be done is in Science 5/13 *Change*, Stages 1 and 2 (see bibliography: 20). Refer back also to the investigation involving capillary flow through fabrics on page 17.

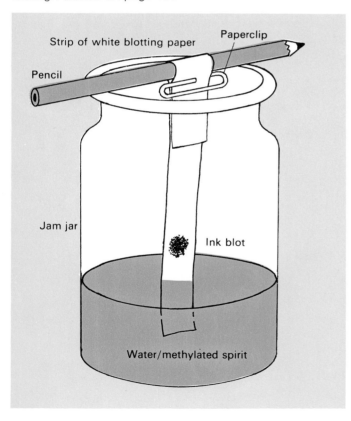

Strip of white blotting paper
Paperclip
Pencil
Jam jar
Ink blot
Water/methylated spirit

Some dyeing methods

Wax resist   You will need:

Pieces of fabric	Candle
Paints	Blotting paper
Dyes, eg Dylon cold	Hot iron
water dye	Newspaper

Inscribe a design on a piece of fabric with the candle. Put the fabric into a solution of paint and dye. Take out the fabric and allow it to dry. Cover it with blotting paper and iron. The heat from the iron will dissolve the wax.

Paste resist   You will need:

Flour and water paste	Paints
Detergent container	Dyes such as Dylon
with cap	Roller
Paint brush	Newspaper
Fabrics	

Use the paste to make patterns and designs on some fabric. You might find it easy to squeeze the paste out of a detergent container. When the paste is dry dye the fabric.

Tie dyeing   Here the fold or the tying resists the dye. You will need:

String and cotton thread	Fabrics
Paints	Pail or large bowl
Dyes such as Dylon	Coins, pebbles or marbles

Fold, tie knots, or tie objects such as stones or coins into some fabric. Immerse the fabric in the dye. Then take it out and allow it to dry.

More detailed information on these methods is available, for instance in *Paints and materials* in this series. See bibliography: 9, 17, 19.

The link with science   The idea behind wax and paste resist is that both wax and paste repel water and therefore also water-based dyes. The investigation on page 16 deals with this principle.

Make a plan for a scheme of work linking resist dyeing

with these investigations. Which would you start with first? Could the two be run side by side?

When you tried one or other method, how successful do you think you were?

Problems
What is the best type of dye to use and how concentrated should it be?

What is the best way of applying dye?

How hard do you have to press the candle in the candle resist method?

How thick should the paste be for paste resist?

How hard do you need to tie the material for tie dyeing and how do you do it to produce a particular effect?

Printing fabrics

You will need:

Paints
Dyes (Helizarin dye and associated chemicals are especially good)
Flat tin lid or a wooden board
Paint brush
Thick pad of newspaper
Jar for the dye
Piece of thin foam rubber in a flat tin
Pieces of white fabric
Wooden cutting board
Penknife or craft knife (careful!)
Clean unpeeled potatoes

Begin by working on odd pieces of discarded fabric. Slice the end off a potato with a sharp knife. Cut a design into the smooth surface. Apply the dye to the surface of the potato. Then apply this surface to the fabric. See the investigations on pages 17, 18.

Problems
What is the best method of making the design?

What is the best way to apply the dye to the potato? With a brush? With an inked pad?

What is the best type of dye to use? You could find out by seeing how 'fast' it is when it is washed in warm soapy water.

See bibliography: 17, 19.

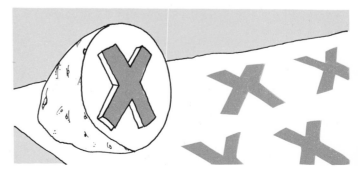

Spinning fibres into thread

You will need:

Raw wool
Detergent
Dowelling (or a pencil)
Plasticine
Carder

Small cup hook
Hand spindle (a proper one
if you can get it)
Craft knife

In spinning, the wool is first carded or combed so that all the fibres lie more or less side by side. The end of the carded wool is then attached to a spindle and as the wool is (slowly) fed into it, it is twisted into thread.

Wash the wool in a warm solution of detergent and water. Rinse and allow to dry, and then comb or card the wool. Cut about 15–20 cm of dowel rod for the spindle. Roll a large lump of Plasticine into a ball; then fix it round the spindle near either the upper or the lower end. This acts as the whorl.

Carding wool

The carded wool can be attached to the spindle in various ways. Some are more satisfactory than others (see diagram). Spin the spindle with one hand while you feed wool in with the other. This might lead to the investigations suggested on pages 24–25.

Problem What effects do changes in the shape, the position and the weight of the whorl have on the action of the spindle?

See bibliography : 11.

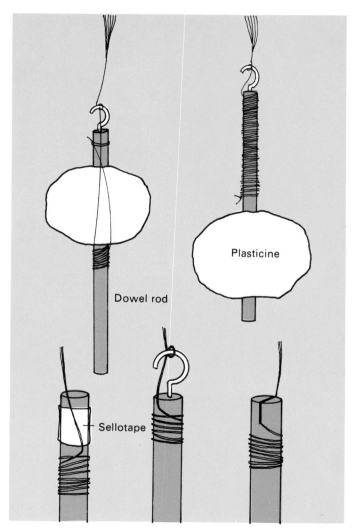

Dowel rod

Plasticine

Sellotape

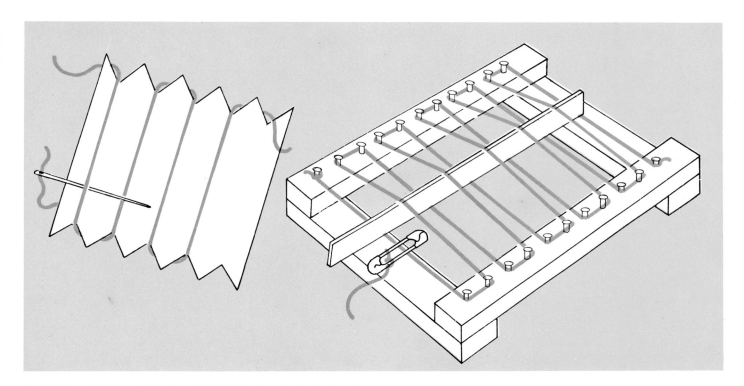

Weaving

You will need:

Piece of fabric with an open weave, eg bandage	Safety pin
Sheet of card	Four lengths of 2 × 1 in wood (say 9 in long)
Scissors	Nails, 1 in and 1½ in
Needle and cotton	1½ in screws
Wool yarn	Bradawl
Card	Screwdriver
Pencil and ruler	Hammer

Examine the open weave fabric. Note the way in which the threads run. The warp threads run the length of the fabric and the weft threads run across them.

Make a simple card loom. Take the sheet of card and make notches along two opposite ends. Wind a long warp thread round the notches; this is the warp. Thick wool makes a good warp thread for children.

Pass the needle and thread alternatively over and under the warp threads. When you have finished one width turn the needle and come back. This is the weft. This makes a simple plain weave. Experiment with other weaves, for example by going over and under pairs of threads.

A simple frame loom can be made by screwing or nailing the four pieces of wood together. One-inch nails are then spaced at about half-inch intervals in a row at opposite ends of the loom. The warp threads are wound between the nails in the same way as for the card loom. Materials like strips of fabric, lengths of thick wool, and even reeds or grasses can form the weft. A safety pin can be used as the shuttle. A ruler inserted through the warp threads and turned so as to lift them up will help threading. This could lead to the investigations on pages 24–25.

See bibliography: 17.

Paper making

The process of turning natural fibre into paper is difficult to carry out in the classroom. As a compromise soft paper can be macerated and remade into sheets. You will need:

Bricks
Water
Two pieces of felt
Wooden frame 50 cm square covered with a wire gauge or wire mesh
Washing-up bowl
Soft toilet paper
Starch boiled in water
Wooden spoon or stick
Wooden board

Beat the paper to a pulp in water. Add the starch solution to make the pulp adhere. Dip the frame in the pulp; shake evenly and allow to drip. Turn out the pulp on to a piece of dampened felt or water-absorbent fabric. Blot, then place another piece of felt on top, then a board on which are placed the bricks. Finally take out the sheet and dry in the air. Refer to the investigation in Chapter 4.

See bibliography: 12, 15.

Keeping and growing

Keeping silkworms Look in *Insects* in the Macdonald Junior Reference Library for information about the silkworm life cycle. *Keeping Animals and Plants in School* by V. B. A. Gregory advises how to keep silkworms. The collection of the silk might lead to the investigation on pages 24 and 25.

See bibliography: 1, 14.

Growing flax and cotton Flax can be grown in the garden. The perennial blue Linum can be obtained from seed merchants. Cotton can be grown in the classroom from seed. See bibliography: 12.

Studying textiles

Geography topics
1 The countries that produce different types of animal and plant fibre.

2 The wool industry in this country: sheep farming.

3 The textile and paper industries in this country. This might lead to the investigations in Chapters 4–7. See bibliography: 3.

Clothes These are a natural starting point for many of the investigations in this book, and topics and questions are easy to think of. For example:

Why do we wear clothes?

Which clothes keep us warm, cool? What are they made of?

Do certain coloured clothes have certain effects on the wearer's comfort?

Are some fabrics more comfortable than others?

Testing the insulation of fabrics with thermometers

History topics

1   The development of spinning and weaving.

2   The development of natural and man-made fibres.

These might lead to the investigations in Chapters 2, 6, 7.

Here is a photograph of part of a children's frieze painting copying the Bayeux Tapestry. The processes of copying and painting would encourage the children to acquaint themselves really familiarly with the weaving techniques of the eleventh century. It would also help them to learn about what clothes people wore then.

See bibliography: 3.

8 Putting ideas together

A checklist

There are many further possibilities. These can be worked out by bringing together ideas that you may have already studied into new combinations. To do this it might be helpful to draw up a checklist based on:

Things Objects or phenomena which the children might usefully investigate.

Activities What the children might do or find out.

Ideas which could be extended or clarified through study of this topic.

Here is a checklist made up from this book. You may wish to add to it. It might be used in conjunction with the bibliography on pages 38–39.

Things

Fibres, natural, man-made	Paper
Thread, yarn	Fabrics
Wax	Parachutes
Fire and flame	Sails
Dirt	Light
Soap and detergents	Paints, inks, dyes
Rope, string	Mordants
Clothes	Prints
Water	Collages
Air, wind	

Activities

Observing
   effects of . . .
   conditions affecting . . .
   products of . . .

Considering
   the uses of . . .
   the history of . . .
Measuring

Comparing
Classifying
Explaining
Experimenting

Printing
Painting
Spinning, weaving, sewing
Dyeing

Ideas

Burning	Shrinkage
Safety	Stretch
Attraction	Elasticity
Repulsion	Breaking strain
Resistance	Weight (mass)
Permeability	Warp, weft
Absorption	Weave
Capillarity	Ply
Crease resistance	Chromatography
Suspension	Wear
Solution	Filtering

From this list you can select particular items, and also combine them, for example *comparing* the *capillary flow* of *water* along different kinds of *string*.

Flow charts

Take one item from the list as your central focus and draw a flow chart of possibilities. This can be used as a basis for discussion with the children and colleagues so as to get their modifications and suggestions. Opposite is an example of a flow chart based on textile materials.

When trying to think of more ideas, which methods or combination of methods did you find most fruitful: the checklist, the bibliography, discussion with children, or discussion with colleagues?

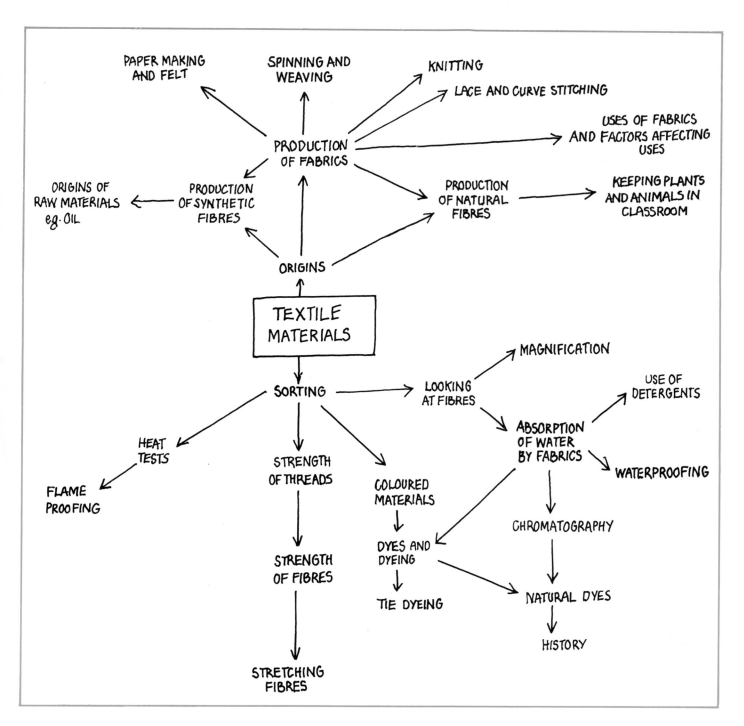

Bibliography

For children to use

1 Cochrane, J. (ed.) (1968) *Insects*. Macdonald Junior Reference Library. Macdonald Educational. See pages 44–46.
2 Cochrane, J. (ed.) (1969) *Textiles*. Macdonald Junior Reference Library. Macdonald Educational. A mine of information on fibres and fabrics. Useful for adults too.
3 Kerrod, R. (1971) *A First Look at Cloth*. Franklin Watts. An account of the various fibres and their origins, and of textile manufacturing processes. Also deals with the history of spinning and weaving. Juniors.
4 Thomson, Ruth (ed.) (1973) *Cloth and Weaving*. Macdonald First Library. Macdonald Educational. Top infants and juniors.

For direct work with children

5 Aston, O. (1972) *Water*. Evans Integrated Themes. See pages 28–29.
6 Bainbridge, J. W., Stockdale, R. W., and Wastnedge, E. R. (1970) *Junior Science Source Book*. Collins. See pages 22–28, 155–160, 161–167.
7 Baker, P. (ed.) (1973) *Science Session Spring 1974*. Teachers' notes and pupils' pamphlet. B.B.C. Publications. See pages 4–7, 12–15.
8 Bird, J. and Diamond, D. (1975) *Candles* (Nuffield/Chelsea College Teaching Primary Science series). Macdonald Educational.
9 Brady, C. (1975) *Paints and materials* (Nuffield/Chelsea College Teaching Primary Science series). Macdonald Educational.
10 B.P. Educational Service (1973) *Science Teaching for ROSLA*. B.P. Educational Service, P.O. Box 5, Wetherby, West Yorkshire. See section 3.1 'How strong are nylon fishing lines?' on page 16.
11 Castino, R. (1975) *Spinning and Dyeing the Natural Way*. Evans. See pages 20–27, 46–52. Tells you all you need to know about spinning, weaving and dyeing. Very useful.
12 Finch, I. E. (1971) *Nature Study and Science* Longman. See sections 74–74/8. An excellent text on teaching primary science with a good section on paper and fabrics.
13 Gawthorpe, L. M. (1966) *Science in the Home: Fabrics and Laundry Work*. Hulton Educational. See pages 8–10, 25–28, 29–32, 43–44. An excellent simple treatment of the topic. Includes lots of useful information on types of fibres, their origins, and the basic manufacturing processes.
14 Gregory, V. B. A. (1972) 'The Keeping of Animals and Plants in Schools'. Natural Science Society Publication No. 11. See page 5.
15 Grigson, Geoffrey (1962) *The Shell Country Book*. Phoenix. Includes a useful account of paper making from natural materials.
16 King, J. and Stewart, R. A. (1970) *Dyes and dyeing*. Griffin Technical Studies. Griffin & George.
17 Melzi, K. (1968) *Art in the Primary School*. 2nd edition. Basil Blackwell. See pages 126–149, 172, 180–183.
18 Piper, B. (1968) *Fibres and Fabrics*. Housecraft Series. Longman. See pages 5–12, 22–25.
19 Pluckrose, H. (1969) *The Art and Craft Book*. Evans. See pages 52–53, 57, 137–141.
20 Schools Council Science 5/13 (1973) *Change*, Stages 1 and 2 and background. Macdonald Educational. See pages 33–36, 45–46, 59.
21 Schools Council Science 5/13 (1973) *Change*, Stage 3. Macdonald Educational. See pages 64–65.
22 Schools Council Science 5/13 (1974) *Children and plastics*, Stages 1 and 2 and background. Macdonald

Educational. See pages 31, 36–43.

23 Schools Council Science 5/13 (1973) *Coloured things*, Stages 1 and 2. Macdonald Educational. See pages 52–56.

24 Schools Council Science 5/13 (1972) *Early experiences*. Macdonald Educational. See pages 57–58, 67–68, 84–88.

25 Schools Council Science 5/13 (1973) *Holes, gaps and cavities*, Stages 1 and 2. Macdonald Educational. See pages 16, 58–60.

26 Schools Council Science 5/13 (1973) *Like and unlike*, Stages 1, 2 and 3. Macdonald Educational. See pages 52–53, 61.

27 Schools Council Science 5/13 (1973) *Ourselves*, Stages 1 and 2. Macdonald Educational. Page 25.

28 Schools Council Science 5/13 (1972) *With objectives in mind*. Macdonald Educational. See pages 21, 43–46.

29 Shepherd, W. (1975) *Textiles*. Finding Out About Science 24. Hart-Davis.

30 Shillinglaw, P. (1972) *Introductory Weaving*. Batsford. See pages 30–31.

31 Sime, I. F. (1970) *Plant Dyes*. Pupils' book and teachers' guide. Ginn Practical Science Projects. Ginn.

For further information and ideas

32 Bassey, M. (1969) *Science and Society*. University of London Press. See pages 37–41. A good simple account.

33 Houvink, I. D. (1975) *Sizing up Science*. John Murray. See page 89.

34 I.C.I. Fibres Limited (1975) 'The World of I.C.I. Fibres'. I.C.I. Fibres Limited, 68 Knightsbridge, London SW1.

35 Lansdown, B., Blackwood, P. E., and Brandwein, P.F. (1971) *Teaching Elementary Science Through Investigation and Colloquium*. Harcourt Brace Jovanovich. See pages 105–107, 119–121, 150–154. (Ask at a library for this book.)

36 Ling, E. M. (1972) *Modern Household Science*. Mills & Boon.

37 Nuffield Secondary Science (1971) *Theme 7 Using Materials*. Longman. See pages 163–184, 199–222. Many good ideas, although the treatment is too advanced for primary school work.

38 Patten, M. (1971) *The Care of Fabrics*. Ginn. See pages 15–22.

39 Taylor, M. A. (1972) *Technology of Textile Properties: an Introduction*. Forbes Publications. A detailed, yet not too advanced, text on the basic science of fibres and fabrics.

Catalogues

40 E. J. Arnold (Art and Craft Catalogue)
Butterley St
Leeds LS10 1AX

41 Dryad Ltd
Northgates
Leicester LE1 4QR
A general catalogue on materials for creative work, including dyes, fabrics, samples of fibres, as well as spindles and carders.

Useful addresses for information

Bowater Corporation Ltd
Bowater House
London SW1

Courtaulds Ltd
Public Relations Department
18 Hanover Square
London W1A 2BB

Flaxspinners and Manufacturers Association of Great
Britain
Public Relations Office
4 Chamber of Commerce Building
Dundee
Scotland

I.C.I. Fibres Ltd
Hookstone Road
Harrogate
Yorkshire

International Wool Secretariat
Education Department
Wool House
6 Carlton Gardens
London SW1

Monsanto Textiles Ltd
Public Relations Department
Monsanto House
10–18 Victoria Street
London SW1

Proctor and Gamble Educational Service
P.O. Box 154, Newgate House
Newcastle-upon-Tyne NE99 1SM
Pamphlet: *Soaps and Soapless Detergents*. A brief but
invaluable leaflet on washing and the use of detergents.

Samples

The following firms will supply samples of materials:

Courtaulds Ltd
Public Relations Department
18 Hanover Square
London W1A 2BB (free)

International Wool Secretariat
Education Department
Wool House
6 Carlton Gardens
London SW1

Monsanto Textiles Ltd
Public Relations Department
Monsanto House
10–18 Victoria Street
London SW1 (free)

National Association of Paper Merchants
35 New Bridge Street
London EC4V 6BH
Will supply a book of various types of paper: quite
expensive but most useful.

Reed International Ltd
Education Services Department
Reed House, 82 Piccadilly
London W1A IEJ

Acknowledgements

The authors and publishers gratefully acknowledge the
help given by:

D. G. Davies, Head of Physics Department, Southlands
College, London SW19
Janet Fairbrother, Gresham Primary School,
Sanderstead, Surrey

The staff and children of:

Arundale School, Pulborough, Sussex
Brookfield County Primary Junior School, Larkfield, Kent
Gresham Primary School, Sanderstead, Surrey

Students and science staff of College of St Mark and
St John, Plymouth, Devon

Illustration credits
Photographs
Kevin Morgan, pages 8, 18
Terry Williams, all others

Line drawings by GWA Design Consultants

Cover design by GWA Design Consultants